Long ago there was a good hobgoblin. His name was Lucky. Lucky wanted to live on a ship.

The skipper of the *Atlantic* said to Lucky, "You can sail on my ship any time!"

"That is fantastic," said Lucky.

Lucky worked hard on the *Atlantic*. He would run up the masts. He would fix the sails.

One night the *Atlantic* was lost. But Lucky could see in the fog.

"I see bright lights," said Lucky.

"You are fantastic," said the skipper.

Lucky was even a good cook. His dinners were excellent!

"This food is fantastic," said the skipper.

Lucky had another important job. He looked for big storms.

"This darkness looks endless!" said the hobgoblin. "There will be a great big storm today."

Once Lucky could see an iceberg.

"That iceberg could harm the *Atlantic*," said Lucky. "I must tell the skipper!"

Lucky saved the *Atlantic*.
The ship did not hit the iceberg.

"You are an important friend," said the skipper.

"This is fantastic!" said the happy hobgoblin.

The End